LIFE ON EARTH

EVERYTHING ABOUT LIFE

DR. MISBAH BHAYO

CONTENTS

PART-01

1.01

How planet Earth was created?

Planet Earth evolved scientifically over billions of years. It began to form roughly 4.5 billion years ago when the solar system was created through a process called accretion, which is the buildup of small particles and objects into larger bodies. Over time, these objects collided and merged together until they reached their current states as planets, asteroids, and other celestial bodies.

In the early stages of formation, Earth was heated by impacts from meteors and comets that brought water and volatile materials to its surface. As temperatures cooled, oceans formed along with landmasses that eventually developed into continents. Over millions of years of plate tectonics activity and erosion from wind and rain, our planet gradually shaped itself into what it looks like today. The continuous evolution of Earth's environment has also allowed for organisms to appear, evolve, and eventually form complex ecosystems. This process continues to shape our planet today and will continue far into the future.

Earth is constantly evolving, with new species and changes in climate occurring every day. It's an ever-dynamic system that scientists are still discovering more about each day. By understanding how Earth developed over billions of years, we can get a better insight into how it functions today and how we can help protect it for future generations.

"The Snowball Earth"

ball Earth is a scientific theory that suggests the Earth once experienced a period of extreme glaciation, or "snowballing." This event occurred around 750 million years ago, when large areas of the planet became covered in ice and snow. The extent of this event is still debated by scientists and may have only affected parts of the planet, rather than the entire surface. During this time, global temperatures fell significantly and much of the ocean was frozen over.

It is believed that Snowball Earth began due to a decrease in atmospheric carbon dioxide levels caused by intense volcanic activity. A drop in sea surface temperatures further exacerbated conditions as cool waters spread across the globe. The ice sheets created reflected sunlight back into space which resulted in even lower temperatures. Eventually, the planet's entire surface was encased in snow and ice.

Though scientists disagree on how long this period of glaciation lasted, there is evidence that Snowball Earth ended when volcanic activity increased atmospheric carbon dioxide levels. This caused global temperatures to rise and allowed for glacial melting. Evidence suggests that the evolution of photosynthetic organisms played a role in the end of the event as they released oxygen into the atmosphere which further warmed the planet. The effects of Snowball Earth can still be seen today in areas with rocky terrain or ancient sedimentary rock deposits.

"Earth Landscape"

The earth's landscape today is due to a combination of various geological processes which have been occurring scientifically since the earliest days of planet formation. Mountains, rivers, and other features were created by the movement of tonic plates and erosion by water wind, and ice.

The of tectonic plates beneath the Earth's surface continually reshaped the surface over millions or billions of years as they collided with one another. This process has formed mountains like the Rocky Mountains in North America and Europe's Alps. Rivers were also formed by this process when two landmasses separated leaving gaps for water to pass through. Erosion was also an important factor in shaping the world's landscapes; weathering and abrasion from strong winds, flowing water, and ice have caused river valleys and canyons to form around the world. Over time, all these processes combined together to create the unique and diverse landscapes we see today.

The scientific understanding of how mountains, rivers and landscape formed on Earth is a reminder of the power of nature's forces that are still at work beneath our feet. It also serves as a fascinating example of how the earth has changed over millions or billions of years in response to geological processes. Knowing more about this process may help us better understand changes occurring now due to climate change so that we can prepare for them and protect our environment more effectively going forward.

1.02

History of life on Earth

Scientists believe that life on earth began some 3.5 billion years ago as single-celled organisms. They hypothesize that these primitive forms of life were created through a process called abiogenesis, which essentially states that inorganic matter combined with energy can create organic molecules and eventually living organisms.

While this is the most widely accepted theory among scientists, it will likely remain largely unproven until further research and experimentation is conducted.

The exact details of how the first life forms came to be are still unknown, but whatever happened has had an incredible impact on our planet's evolution since then. From those humble beginnings, we've seen species multiply, become more complex, and spread all over the globe. Today, there are an estimated 8 million different species of living things, making life on Earth one of the most diverse and fascinating phenomena in our universe.

We may never fully comprehend how these first organisms came to be, but their contribution to the history of life on Earth is undeniable. From those tiny creatures that existed billions of years ago, we now have a planet teeming with life-spanning from microscopic organisms like bacteria to massive mammals like blue whales. The story of life on Earth is an incredible journey, and it's all thanks to those very first living beings who started it all.

What is Life?

Life is a term that has been used to describe various phenomena in the universe. Scientifically, life can be defined as living organisms that have the ability to grow, reproduce, and interact with their environment. This definition applies to both single-celled organisms like bacteria and complex multi-cellular organisms such as humans. Life is thought to have evolved on Earth over billions of years and is an integral part of the planet's ecosystem. Life forms are also found in other parts of the universe, such as on other planets or moons, although none that have been discovered so far are known to be capable of reproduction. As we continue to explore the universe, it is possible that life may be discovered in other places. In any case, life on Earth is an incredible miracle that deserves recognition and celebration.

The study of life is the field of biology, which involves understanding the structure and behavior of living organisms. This includes studying their anatomy, physiology, genetics, evolution, ecology and many other topics related to life. Biology has made incredible advances in recent years, from uncovering the genetic code to discovering new species. Through this research we have been able to gain valuable insight into the workings of life on Earth and how we can better protect it.

Understanding what is life and appreciating its complexity will help us further appreciate the incredible gift that living organisms provide us with, from their essential role in maintaining the health of our ecosystems to providing us with food and other resources.

1.03

First animal on Earth

Scientifically, the first living animal to appear on Earth is hard to pinpoint. Generally speaking, scientists believe that animals evolved from single-celled organisms between 600 and 700 million years ago. The earliest known forms of life were simple organisms such as bacteria, algae and protozoans. These primitive creatures eventually evolved into more complex species such as jellyfish, worms and starfish.

Eventually, these aquatic animals shifted onto land where they eventually evolved into the many different types of terrestrial animals we see today. Thus, although it's impossible to determine which one was truly the first living animal on earth, it is clear that evolution has played a key role in bringing us to where we are today.

From this evolutionary process, we can infer that the first living animal on Earth was most likely some kind of single-celled organism or primitive aquatic creature. Although no clear scientific evidence exists to settle this debate, it is safe to say that the history of animals and their origins is an intriguing and fascinating topic worthy of further exploration.

"The simple sea sponge"

It is scientifically accepted that the first animal on Earth was the simple sea sponge. First appearing in the fossil record around 650 million years ago, it's believed that these early sponges helped to create an oxygen-rich ocean environment which allowed for other forms of life to thrive.

The simple sea sponge genus is still one of the most abundant and diverse groups of animals today, with over 15,000 different species living in oceans around the world. This shows just how successful this ancient group has been in adapting and surviving throughout their long evolutionary history. Despite their primitive appearance, modern sponges are complex creatures capable of incredible feats like creating vast coral reefs, filtering huge volumes of water, and even providing a home for many other marine animals. So, the next time you go to the beach and take a dip in the ocean, keep an eye out for these amazing creatures and remember that they have been around since long before we were here.

In conclusion, it is accepted by scientists that the first animal on Earth was the simple sea sponge. This ancient species has proven to be incredibly successful throughout its long evolutionary history, with many of their descendants still thriving today. Not only are modern sponges fascinatingly complex creatures able to perform astounding feats, but they can also teach us a lot about our own evolution over time.

1.04

First land plants plunge on Earth

The first land plants to plunge on Earth were scientifically known as the Devonian period, which was geographically located in an area referred to as the "Devonian Savannah" - an extensive area of lowland where many species of early terrestrial vegetation flourished. This period marked a huge evolutionary jump for plants; from largely aquatic forms that had survived since before the Cambrian explosion, to some of the earliest ancestors of today's vascular plants. During this time, plant diversity increased dramatically and photosynthetic capabilities allowed them to take advantage of more sunlight than their predecessors. The Devonian period saw an incredible adaptation process, with several new types of leaves and root systems emerging. This in turn led to a surge in oxygen production in the atmosphere which enabled other lifeforms to emerge. This period is considered one of the most important evolutionary stages in the history of Earth, and resulted in a huge diversity of plants that are still around today. This evolutionary jump was an essential step for the development of terrestrial ecosystems, and it laid the foundations for life as we know it today. The Devonian period is sometimes referred to as the "Age of Fishes" due to the wide array of fish species that evolved during this time. These fishes were the first vertebrates to inhabit land and provided a food source for other animals. In addition to pioneering plant life, this period also saw a diversification in animal life, especially arthropods like insects and spiders. It's estimated that by the end of the Devonian period there were around 20,000 different species on Earth, many of which still exist today

"Moss versus Rocks"

Mosses have been around for a long time, and although they are much smaller than rocks, they have an important role in the Earth's ever-evolving landscape.

Scientifically speaking, moss is an ancient plant species that dates back over 400 million years ago. Moss reproduces by spores, not seeds like other plants, and differs from other plants in its lack of flowers and woody tissue. Geographically speaking, moss is one of the most widely distributed plants on Earth since it can survive in both cold and warm climates as well as all types of terrain—from arctic tundra to tropical rain forests. They are also extremely resilient plants with a fast growth rate; some varieties can even grow up to 20 cm per year! Mosses are also incredibly important in the environment, acting as natural water filters and providing shelter to both humans and wildlife. They are also a food source for many organisms, including insects, birds, small mammals, reptiles, and amphibians. So, while rocks have been around since the beginning of time, moss has its part to play in shaping our world. Its enduring presence over millions of years proves that it is more than just an insignificant organism—it's an integral part of our planet's ecology.

In conclusion, while rocks are much larger and more durable than moss, their presence in the Earth's landscape is no match for the long-term impact of moss. Moss has been around since before recorded history and continues to play an important role in our environment today. So, although they may not be as grand or imposing as rocks, their unique characteristics make them a vital part of our planet's natural history.

"Etiological succession"

Etiological succession of plants is a process that occurs naturally and is scientifically described as the chronological replacement of one plant species by another in an area over time. This type of succession often takes place geographically, occurring differently depending on the geographical location and climate. Etiological succession can occur due to many factors, such as changes in soil fertility or climatic conditions, natural disasters, human activities, or other disturbances. As new species arrive and compete with existing ones for resources, some will become dominant and eventually replace others over time. This process can take many years to reach its full potential, but it is a crucial part of maintaining healthy ecosystems. By studying etiological succession of plants, we can better understand how to conserve our planet's biodiversity in the long-term. By doing so, we can ensure that ecosystems are able to remain stable and productive over time, regardless of any changes in their environment. This will ultimately benefit both humans and wildlife alike.

Etiological succession of plants has been studied by scientists for many years, and its mechanisms are still not fully understood. However, it is clear that this natural process plays an important role in keeping our planet's biodiversity healthy and functioning properly. For example, by understanding etiological succession of plants, we can help restore areas that have been damaged due to human actions or natural disasters. We can also gain insight into how certain species might respond to climate change and other environmental shifts in the future.

1.05

The Jurassic periods

The Jurassic Period is a geological period that took place between 201.3 and 145 million years ago, during the Mesozoic Era. It was an era of great biological diversification where many new species, including birds and mammals, evolved from their ancient reptilian ancestors. The Earth's climate during this time remained generally warm and tropical throughout most of the period.

The periodic volcanic eruptions and tectonic shifts caused changes in sea levels, which resulted in the creation of various land bridges connecting different parts of the world.

In terms of geology, Jurassic rocks are typically classified as either limestone or sandstone due to their widespread presence throughout Europe, North America, South America, Asia, Africa and Australia. During this time some of the most famous species of dinosaurs, including the stegosaurus, velociraptor and triceratops, lived and evolved. The Jurassic period is also noteworthy for being the time during which flowering plants first appeared.

The fossil record from this era reveals a wide variety of animal life as well as an abundance of evidence relating to ancient climates. Studies of these fossils have allowed scientists to gain a better understanding of how different species evolved over time and what kind of conditions they were adapted to survive in. This has provided invaluable insight into the workings of the prehistoric world, affording us greater knowledge about our planet's scientific past.

Overall, The Jurassic Period is one that can be studied scientifically with great interest due to its wealth of information and vast array of fossils. It was an era that saw the evolution of many species and served as a catalyst for the evolution of many more in later eras. It is an integral part of Earth's history and its importance should not be underestimated.

Overall, The Jurassic Period is one that can be studied scientifically with great interest due to its wealth of information and vast array of fossils. Fossils found from this period provide invaluable insight into how animals evolved over time, what kinds of conditions they were adapted to survive in, and how different climates affected life on earth at the time. Through research on these ancient specimens scientists have been able to gain a greater understanding of our planet's scientific past, making it one of the most important periods in Earth's history.

Overall, The Jurassic Period is one of the most important eras in scientific history due to its abundant fossil record and wealth of information it provides. Through careful research on these specimens' scientists are able to gain a greater understanding of how different species evolved over time, what kind of conditions they were adapted to survive in and how climate changes affected life at the period. This knowledge has enabled us to develop a better understanding of our planet's scientific past and its importance should not be underestimated. With this insight we can continue to further explore and discover the wonders that lie within our world's geological past.

"Mesozoic Era – The age of Dinosaurs"

The Mesozoic Era, which lasted from about 252 million to 66 million years ago, is known as the age of dinosaurs. Scientifically speaking, this era marks the time in Earth's history when reptiles dominated land life and some species of dinosaurs evolved for the first time.

During this vast stretch of time, Earth underwent major changes in its climate and geography. The Mesozoic Era was divided into three periods: Triassic (252-201 million years ago), Jurassic (201-145 million years ago) and Cretaceous (145-66 million years ago). Each period had a distinct set of characteristics that affected the evolution of plant and animal life during that epoch. The amazing creatures we think of today as "dinosaurs" evolved during the Mesozoic Era and went extinct at the end of the Cretaceous period. Although most dinosaurs were large, there were many varieties ranging from small carnivores to giant herbivores. Dinosaurs left behind abundant fossil records that provide us with an incredible window into their world. Through this evidence, scientists have been able to learn a great deal about how these creatures lived, interacted with one another, and eventually died out. It is important to remember that although dinosaurs are long gone, they still hold an important place in our understanding of

Earth's ancient history and evolution. The Mesozoic era was a time when fantastic creatures roamed the land and changed the course of Earth's future forever.

"Early mammals"

Early mammals began appearing on land during the Paleocene epoch, which was part of the Cenozoic era. These mammals were scientifically classified as being from a group called Therapsida, which included both mammal-like animals and reptiles descended from creatures that lived in the water.

Their legs developed to become increasingly suited to running on land, while their backbones became less flexible in order to provide more efficient support for walking. Additionally, their teeth and jaws changed to enable them to eat food found on land - mainly plants and small insects. This adaptation enabled them to survive in an environment where they weren't constantly threatened by predation. Over time, some of these early mammals evolved into larger species with different diets, such as omnivores or carnivores. By the Oligocene epoch, many of these species had become extinct while others had evolved into specialized species that were better adapted to their environment and more competitive with other animals in the same ecological niche. This evolutionary process is still ongoing today, as mammals continue to adapt to changing environmental conditions on Earth.

Overall, early mammals began to appear on land in the Paleocene epoch as part of the Cenozoic era and have since undergone a process of specialization and adaptation in order to survive and thrive in various environments. Furthermore, through continued evolution, they are still changing today to best suit their current ecological niche.

"Discovery of Juramaia"

Juramaia sinica is a scientifically important species, as it is believed to be the earliest known ancestor of mammals. Discovered in China's Liaoning Province, Juramaia lived more than 160 million years ago during the Jurassic Period and provides valuable insight into the evolution of mammals.

It was discovered by a joint team of researchers from China and the United States using advanced fossil-collecting and imaging techniques.

The discovery of Juramaia helps to fill in many gaps about mammal evolution that were previously unknown or poorly understood before this discovery. For example, it reveals that mammals had already split off into different branches on the evolutionary tree over 160 million years ago - much earlier than previously thought - indicating an incredibly long evolutionary history for mammals.

Moreover, its anatomy also shows that it had features of both primitive and early mammalian species, such as specialized teeth for carnivorous feeding and a well-developed middle ear with an integrated system of hearing structures. These characteristics further support the hypothesis that Juramaia was an early ancestor of modern in mammals.

The discovery of Juramaia is thus an important step forward in our understanding of mammal evolution, unlocking many new possibilities for research into this fascinating topic. It is a testament to the power of scientific investigation to uncover new insights about our world and provide us with a greater knowledge about the past. As such, it will surely remain a landmark discovery for many years to come.

"Human Born"

Scientifically, it is believed that humans were first born on Earth approximately 200,000 years ago. This is based on evidence from fossils, molecular genetics and archaeological records. The earliest human-like species evolved in Africa around 6 million years ago during the late Miocene epoch.

Homo sapiens (modern humans) are believed to have descended from a common ancestor shared with Neanderthals and Denisovans about 400,000 years ago. Through migration and interbreeding the human species spread across the world, eventually leading to the emergence of distinct populations which further adapted through evolution to their local environment over time.

Thus, each population developed its own genetic profile and physical characteristics such as skin color and facial features. Ultimately this resulted in what we see today as the diverse range of humans around the world. It is also believed that early humans had interacted with other hominid species. Interbreeding may have occurred between Homo sapiens, Neanderthals and Denisovans which led to the exchange of genetic material and shared traits among these species. This is evident in modern human DNA which contains a small amount of Neanderthal DNA due to ancient interbreeding.

Thus, it appears that modern humans today are the result of an evolutionary process that has spanned millions of years. From a scientific perspective, this is how humans were first born on Earth.

It is important to note that this is not the only accepted theory of human evolution. There are some scientists who believe that humans were created by an omnipotent being, while other theories exist as well. Ultimately, there is no single answer to this question and much of it depends on one's own personal beliefs and worldviews. Nonetheless, the scientific evidence suggests that humans evolved gradually over millions of years from a common ancestor shared with Neanderthals and Denisovans. This appears to be the most accepted explanation for how we came to be.

1.06

History of stone age

The Stone Age is a period in human prehistory when early humans used mainly stone tools. It is generally considered to have begun around 2.6 million years ago and ended between 3,000 and 4,000 years ago. This time period is marked by significant technological advances and the emergence of Homo sapiens as the dominant species on Earth.

Scientifically speaking, the Stone Age can be divided into three distinct periods: Lower Paleolithic, Middle Paleolithic, and Upper Paleolithic. The Lower Paleolithic began around 2.6 million years ago with primitive stone tools such as hand axes being developed by hominins (early humans). During this period, early humans also began using fire for cooking and for protection from predators. Over time, these tools became more sophisticated and efficient as early humans learned to control fire and hunt larger prey.

The Middle Paleolithic period lasted from 300,000 years ago to 30,000 years ago and was marked by the development of complex tools such as blades and scrapers. During this period, humans also developed religious beliefs and art forms such as cave paintings.

Lastly, the Upper Paleolithic period began around 30,000 years ago and lasted until around 10,000 BC. This period is associated with the development of advanced stone tools and weapons as well as pottery-making techniques. It is during this time that Homo sapiens truly flourished due to their superior technology compared to earlier hominins.

Overall, the Stone Age was a period of significant technological advancement and cultural development that paved the way for modern humans. It is during this time that humankind began to form complex societies and create works of art that still inspire awe today. As such, it is an important part of human history that deserves to be studied and understood.

The Stone Age may have ended thousands of years ago, but its legacy lives on in our technology, culture, and art. When we look back at this important period in history, we can gain a deeper understanding of how far humanity has come since then and where we might go next.

"The three Age system"

Three Age System, which divides prehistory into the Stone Age, Bronze Age and Iron Age, is so important. It helps us to better understand our past and how we got to where we are today.

Overall, The Three Age system allows us to study and better understand the technological advancements that took place during the Stone Age as well as its lasting legacy on modern humans. By studying this period in history, we can gain a deeper appreciation for how far humankind has come since then. In addition, we can develop a greater understanding of how cultures evolved over time and what motivated early humans to create works of art that still inspire awe today. Through this knowledge, we can gain insights into our own development as a species and what the future may hold. Therefore, it is essential that we continue to study and understand the Three Age System so that our past can help inform our future.

"Prehistory"

During the Stone Age, humans lived scientifically by relying on their observation and understanding of nature to survive. They used the resources available to them such as rocks, bones, and antlers to craft tools like spears and axes that allowed them to hunt animals, gather food, and create shelter. They also incorporated fire into their daily lives for warmth and cooking purposes. This period of human history was largely focused on learning how to manipulate the environment around them in order to survive. As advances in technology emerged over time, humans were able to build more efficient tools which enabled them to further explore their environment and expand upon their knowledge base. Eventually, this led humanity into a new era with more complex societies that had the capacity for technological advancement. The Stone Age is an important part of humanity's history, as it laid the foundation for many of the modern conveniences that we take for granted today.

By looking back at our prehistory, we can gain a better understanding of how far humans have come in terms of technology, society, and culture. This knowledge can also help us to appreciate what we have now and to look forward to greater achievements still ahead. The Stone Age was a time when humans had to work hard for their survival and use their creativity and ingenuity to find solutions. It is a reminder that even in times of difficulty, we can still make progress if we put in the effort.

How people of (Stone Age) make the stone crockery items?

In ancient times, people living in the Stone Age made their own stone crockery items. The first step was to collect raw materials from the surroundings such stones, clay and natural fibers.

The next step was to shape the material into a desired form. This could be achieved by pecking, chipping or abrading it with tools available at that time. Abrasive tools were usually made of flint which is a hard sedimentary rock used for cutting and grinding grainy surfaces. Stones like quartzite, limestone or basalt were also used for making these tools. Depending on what kind of item was being created - bowls, plates or cups - different techniques were employed for shaping them out of stone.

PART-02

2.01

Cell Biology

Cell and Tissues.

Cell and tissues are the building blocks of all living organisms. Cells are the smallest unit of life, and they form tissues when many cells group together to form a specific structure or organ. Cells vary in size and shape, but most have a nucleus which acts as the control center for the cell. The nucleus contains genetic information that determines how a cell develops, grows and interacts with its environment. In addition to the nucleus, cells also contain cytoplasm, membrane-bound organelles, such as mitochondria and ribosomes, as well as other molecules like proteins and enzymes. Together, these components allow cells to perform their necessary functions so that an organism can stay alive. Tissues are groups of cells with similar features and functions that work together to perform specific tasks. For example, muscle tissue is made up of many muscle cells that provide strength and movement while skin tissue is composed of epithelial cells which form a barrier between the organism's internal environment and its external environment. These tissues are essential for an organism to survive, as they allow them to interact with the world around them and maintain homeostasis.

"Principles Of Microscopy"

In order to observe and study cells and tissues, scientists use a tool called the microscope. The microscope was first invented by Antoni van Leeuwenhoek in 1674 and it has continued to play an important role in biology ever since. Microscopy is the science of using microscopes to observe objects that are too small to be seen with the naked eye. It enables researchers to magnify objects many times their original size so that they can better understand them.

The principles of microscopy involve three main steps: illumination, magnification, and resolution. Illumination provides light for the specimen being studied which allows it to be viewed under brightfield or darkfield conditions depending on if light passes through or is reflected from the specimen, respectively. Magnification is the process of enlarging an object such that details can be viewed and studied more easily. Resolution is the ability to distinguish between two objects at a distance, and it is achieved through power or numerical aperture (NA). The higher the NA, the higher the resolution and therefore the clearer image that can be seen.

By following these steps, scientists are able to observe cells and tissues in great detail. This has allowed biologists to gain a better understanding of life at its most basic level which has had far-reaching implications for science and medicine alike. Microscopy has been instrumental in making major discoveries such as discovering DNA, understanding how genes work, identifying bacteria and viruses, treating diseases like cancer, as well as many other medical breakthroughs.

"The Light Microscope"

A light microscope is an instrument of science used to observe small objects that cannot be seen with the naked eye. It works by using a lens or series of lenses to magnify the image of an object, often on a transparent plate called a slide. Light microscopes can be used to examine specimens as small as one micrometer in size. These microscopes are particularly important for studying biological samples such as cells, bacteria and other microorganisms, allowing researchers to gain insight into their structure and behavior. Additionally, they are commonly used in educational settings to teach students about biological concepts at the cellular level. Higher power light microscopes are also capable of producing highly detailed images of much larger objects, making them invaluable tools for scientific research. By combining different lenses, filters and imaging technologies, light microscopes can be used to observe a wide variety of objects in amazing detail. As this powerful instrument continues to evolve with new technological advancements, it will remain an invaluable tool for researchers around the world.

"The Electron Microscope"

An electron microscope is a type of microscope that uses electrons to create an image of an object. It works by passing a beam of electrons through the object, and then detecting the scattered electrons. The pattern created from this process is used to construct an image. This type of microscope allows for higher magnification than traditional optical microscopes, and can be used to view tiny features that would otherwise not be seen with a regular microscope. Electron microscopes are commonly used in industrial settings, medical research, and materials science. They can also be used to investigate historical artifacts or other objects on a microscopic level. Additionally, electron microscopes are capable of producing 3-

dimensional images electron as microscope well is as a providing type information of about microscope the used chemical to composition observe of objects certain too materials small. to In be short seen, by the optical naked eye. electron microscope It is works capable by of passing magnifying a objects beam up to of the electrons 100,000x through magnification the range object. The and pattern electrons created scattered from off this the process object is are used then to detected construct and an analyzed image. This for type microscopic of details. microscope Electron allows microscopes for are higher commonly magnification used than in traditional industrial optical settings, microbiology laboratories, materials science, medical research, and forensic analysis. They can also be used to investigate historical artifacts or other objects on a microscopic level. Additionally, electron microscopes are capable of producing 3-dimensional images as well as providing information about the chemical composition of certain materials too small to be seen by the naked eye.

"Prokaryotic And Eukaryotic Cells"

Prokaryotic cells are the simplest type of cell and lack a membrane-bound nucleus. They typically have one circular chromosome made up of DNA that is located in the nucleoid region. Prokaryotes include bacteria, cyanobacteria, and archaea. These cells lack organelles such as mitochondria, chloroplasts, and Golgi apparatus which are found in eukaryotic cells.

Eukaryotic cells are more complex than prokaryotic ones and contain a nucleus surrounded by a nuclear membrane. The chromosomes within the nucleus also contain linear strands of DNA rather than just one circular chromosome like in prokaryotic cells. Eukaryotes can range from single-celled organisms like protists to multicellular species like animals and plants. They contain organelles such as mitochondria, chloroplasts, Golgi apparatus, endoplasmic reticulum, and lysosomes which are not found in prokaryotic cells. Additionally, eukaryotic cells have a cytoskeleton that helps to provide structure and movement.

In summary, the main differences between prokaryotic and eukaryotic cells are their genetic material organization and the presence of organelles. Prokaryotes have one circular chromosome while eukaryotes have linear chromosomes located within a nucleus surrounded by a nuclear membrane. Additionally, eukaryotes have organelles such as mitochondria, chloroplasts, Golgi apparatus, endoplasmic reticulum, and lysosomes while prokaryotes do not.

"Plant Cell"

A plant cell is a type of eukaryotic cell that makes up the fundamental building blocks of all plants. Plant cells have several distinct characteristics and properties compared to other types of cells. They are larger than animal cells, have rigid walls composed of cellulose, contain chloroplasts which allow them to perform photosynthesis, and store special chemicals called vacuoles for protection and energy storage. The nucleus controls all activities within the cell, while mitochondria generate energy for the cell from nutrients. Additionally, plant cells contain microtubules responsible for maintaining shape and structure, as well as endoplasmic reticulum which produces proteins used in cellular processes. These parts work together to ensure the successful functioning and growth of plants on both cellular and organismal levels.

Additionally, plant cells have specialized structures that enable them to sense their environment through the production of hormones and other chemicals which allow plants to respond to environmental changes such as light or water availability in order to survive and thrive in various conditions and habitats. Furthermore, plant cells are capable of dividing rapidly, a process called mitosis, which allows for growth and repair within the organism's body over time and ensures the species' long-term survival in changing environments. This unique combination of properties makes plant cells essential components of all living plants on Earth today.

"Reproductive System of Plants"

The reproduction system of plant cells is a complex process that involves the development and growth of new cells. Plant cells divide in two different ways: mitosis and meiosis. Mitosis occurs when a cell divides into two identical daughter cells, with each one having the same genetic information as the original parent cell. This type of division usually takes place during normal growth or repair processes. On the other hand, meiosis occurs when a cell divides into four genetically unique daughter cells, which are used for sexual reproduction in some plants. During this process, chromosomes from each parent pass on to their offspring so that they can form distinct traits. Both mitosis and meiosis play an essential role in maintaining healthy plant populations by ensuring genetic diversity amongst individuals within a species. As a result, this is an important part of the reproductive system of plant cells and aids in the survival of plants in changing environments.

In addition to mitosis and meiosis, pollen grains are also used in plant reproduction. Pollen contains genetic material from one parent, either male or female, which then combines with pollen from another individual to form a zygote (fertilized egg). The zygote develops into an embryo and eventually grows into the mature plant, containing both parental genes. Therefore, to reproduce successfully plants must have their male and female parts functioning properly so that successful pollination can occur. This is why it is essential for plants to have functional reproductive systems such as flowers with both male and female parts.

In conclusion, the reproduction system of plant cells is a complex process that involves mitosis, meiosis, and the development of genetic material from each parent in order to produce offspring with unique traits. This reproductive system plays an important role in preserving species by ensuring genetic diversity amongst individuals within a population. Additionally, it is essential for successful pollination so that plants can reproduce successfully. With this knowledge we can better understand how plants maintain their populations and why they are such resilient organisms.

What is Tissue?

Tissue is generally defined as a cellular organizational level intermediate between cells and a complete organism. It is an ensemble of similar cells from the same origin that together carry out a specific function. In its simplest form, tissue is composed of cells that are embedded in a matrix of proteins and other molecules such as proteoglycans, collagen, and elastin.

In summary, there are four main types of tissues in animals: epithelial, connective, muscular, and nervous. Each type has distinct functions and properties which enable the body to perform its necessary functions. Additionally, tissues are composed of cells embedded in an extracellular matrix composed of proteins and polysaccharides. Understanding tissue types is important for understanding how different organs and systems in the body function.

Epithelial Tissue: Epithelial tissues are made up of layers of closely packed cells with little intercellular material between them. These layers provide protection to underlying tissues and organs, as well as line body cavities and form glands. Examples of epithelial tissue include the skin and lining of the digestive tract.

Connective Tissue: Connective tissue is the most abundant type of tissue in animals. It serves to connect different structures within the body, and consists of cells embedded in an extracellular matrix composed of proteins (i.e., collagen) and polysaccharides (i.e., glycosaminoglycans). Connective tissue can be divided into three categories: dense connective tissue (such as tendons), loose connective tissue (such as adipose), and specialized connective tissues (such as cartilage).

Muscular Tissue: Muscular tissue is responsible for moving organs and other body parts. It can be divided into three categories: skeletal muscle, which is attached to bones and produces movement; cardiac muscle, which is found in the walls of the heart and pumps blood; and smooth muscle, which is found in the walls of hollow organs and produces peristalsis.

Nervous Tissue: Nervous tissue consists of cells called neurons that transmit information throughout the body via electrical impulses. Neurons are generally organized into different structures such as nerves, ganglia, or brain tissue. The main components of nervous tissue are nerve cells (neurons) and glial cells (which provide support).

2.02

DNA (Deoxyribonucleic Acid)

DNA (deoxyribonucleic acid) is a molecule that carries genetic information in all living organisms. It is made up of two strands of nucleotides wound together like a double helix, and held together by weak bonds between the bases on each strand. DNA contains genetic instructions from our ancestors that are passed down to us through generation after generation.

In the 1950s, scientists discovered that DNA was responsible for carrying genetic information within living organisms. This discovery has revolutionized modern genetics and biology, allowing scientists to study disease-causing genes, develop treatments for human health problems, and even modify crops to be more resistant to pests or drought.

DNA consists of four different kinds of molecules called nucleotides: adenine (A), thymine (T), guanine (G) and cytosine (C). These nucleotides are arranged in a specific sequence to form strands of DNA. Each strand is connected to its complementary strand by hydrogen bonds between the bases. The two strands together form a double helix structure, with adenine on one side connecting with thymine and guanine connecting with cytosine on the other side. The code contained within the DNA dictates how proteins are made, which control all cellular activities from eye color to metabolism. This code is composed of a genetic language called ATCG, or "base pairs", that tells cells which amino acids they should assemble into proteins.

A Brief History of (DNA)

DNA (deoxyribonucleic acid) is the hereditary material found in all living organisms. It was first identified in 1869 by a Swiss physician and biochemist named Friedrich Miescher, who discovered that white blood cells from pus had an acidic substance he called nuclein. He later realized that this acidic substance contained a phosphorous-containing sugar and nitrogenous bases, which would come to form the basis of our understanding of DNA today.

In 1944, Oswald T. Avery and his colleagues at Rockefeller University began to explore the implications of genetic material being composed of DNA rather than proteins as previously believed. They published their results after demonstrating that it was indeed DNA carrying genetic information when they transferred a virulent strain of pneumococcus bacteria to a non-virulent strain.

In 1952, American biochemists James Watson and Francis Crick presented their double helix structure of DNA, for which they later won the Nobel Prize in 1962. This model was based on the X-ray crystallographic images of DNA by British physicist Rosalind Franklin. The model proposed that DNA consisted of two strands wound around each other like a twisted ladder, with four nitrogenous bases pairing up along the length of the molecule (adenine with thymine and guanine with cytosine). The discovery showed how genetic information is encoded, replicated, and transmitted and provided insight into how mutations can occur in individuals.

This breakthrough revolutionized our understanding of genetics and has been fundamental in developing modern medicine, including the development of vaccines, treatments for genetic diseases, and new methods for diagnosing and preventing disease.

Today, our understanding of DNA is still evolving as scientists work to unlock its mysteries. We now have powerful tools like sequencing that allow us to sequence entire genomes at a fraction of the time and cost it took just a few decades ago. This technology has enabled us to explore evolutionary relationships between species, understand how traits are inherited from one generation to the next, develop improved medical treatments based on our genetic makeup, and even uncover ancient secrets locked within ancestral DNA

(DNA) Structure.

DNA, or deoxyribonucleic acid, is a complex molecule that carries the genetic information for all living organisms. It consists of two strands wound around each other in a double helix formation and contains four different nucleotides: adenine (A), thymine (T), guanine (G), and cytosine (C). These nucleotides are arranged in specific combinations which form the code for life.

The DNA double helix is composed of two sugar-phosphate backbones, with each sugar connected to one of the four bases. The sequence of these bases determines the genetic information stored in its structure; A always pairs with T, while C always pairs with G. This arrangement is known as base pairing.

The DNA double helix is highly stable, allowing it to remain intact for many years. This stability is due to the hydrogen bonds between the two strands of DNA, and the fact that chemical reactions only occur when there are specific combinations of bases present in the sequence. The strength of these bonds also ensures that there is no significant variation in DNA sequences over time, making it possible for living organisms to pass genetic information on from generation to generation with high accuracy.

DNA has several unique properties which make it an ideal molecule for storing genetic information. Firstly, its structure allows it to efficiently store large amounts of data; a single strand of DNA can contain up to three billion nucleotides. Furthermore, its ability to replicate itself so accurately ensures that genetic information is not lost or distorted over time, making it ideal for passing on vital information from one generation to the next.

In addition to its structural stability, DNA also has the incredible ability to self-repair when damaged. This means that if DNA strands become broken or mutated due to external factors such as radiation or chemical exposure, cells can repair the damage by finding and replacing the incorrect base pairs with their correct counterparts.

Overall, the structure of DNA is a remarkable example of nature's ingenuity; its combination of stability and accuracy enable it to store complex genetic information which can be passed down through generations without significant loss or alteration. As such, DNA serves as an essential component in all living organisms, and will likely remain a vital part of the biological world for many years to come.

The unique structure of DNA makes it an essential component in life, and its ability to accurately store genetic information makes it invaluable for research into evolution, disease prevention, and much more. Its stability and self-repair capabilities ensure that its data remains intact over long periods of time, making it the ideal molecule for preserving life's secrets.

2.03

Ribonucleic Acid (RNA)

RNA, or Ribonucleic Acid, is a type of nucleic acid that carries instructions from DNA for controlling cell functions. It is directly involved in the protein synthesis process and provides a template for the production of proteins. RNA consists of ribose sugar, phosphate groups, and nitrogenous bases adenine (A), guanine (G), cytosine (C) and uracil (U). Compared to DNA, RNA has a single stranded structure and is much shorter.

RNA can be divided into three main types: messenger RNA (mRNA), transfer RNA (tRNA) and ribosomal RNA (rRNA). mRNA serves as an interpreter between DNA's code and the translation machinery needed for protein synthesis. It is transcribed from a gene in the nucleus and acts as an intermediary between DNA and proteins. tRNA helps to shuttle specific amino acids to ribosomes during protein synthesis. It has a cloverleaf-like structure with several loops, including the anticodon loop used to recognize codons on mRNA and link them to their corresponding amino acid. Finally, rRNA resides within ribosomes where it assists in decoding mRNA into proteins. Together, these three types of RNA form the basis for any biological process that relies on protein synthesis.

In summary, RNA is an important molecule involved in many cellular processes such as transcription, translation and replication. Its three main types all play crucial roles in translating genetic information into proteins for carrying out various cellular functions. RNA is an essential component of all living organisms and without it, life as we know it would not exist.

The study of RNA is called ribonucleic acid research (or RNAR). It uses techniques such as polymerase chain reaction (PCR) and cloning to understand how the molecules interact with one another and their roles in gene expression. The use of bioinformatics software also helps researchers analyze large data sets related to RNA structure and function. Through these methods, scientists are able to gain insights into how changes in RNA sequence or structure can lead to diseases and other medical conditions. Additionally, RNAR has applications in biotechnology, allowing researchers to produce custom proteins by manipulating the genes that encode for them. RNAR continues to be an active field of research, providing new and exciting discoveries about this essential molecule.

A Brief History of (RNA)

RNA, or ribonucleic acid, is a complex molecule present in all living things and plays a critical role in the regulation of gene expression. In 1886, Richard Altman first identified RNA as a distinct cellular component. However, it wasn't until the 1890s that scientists began to understand its biological importance.

In 1902, German biochemist Phoebus Levene proposed that RNA was composed of four types of monomer units: adenine (A), cytosine (C), guanine (G) and uracil (U). The genetic code of life was finally uncovered in 1961 by Marshall Nirenberg and Heinrich Matthaei. They showed that UAC codes for phenylalanine, the first of 64 codons discovered.

In 1965, Francis Crick proposed that genetic information is stored in DNA and transcribed into RNA to make proteins. This discovery laid the foundation for our modern understanding of how molecular processes are regulated in cells.

Since then, our knowledge about the importance of precise regulation of gene expression has grown exponentially. In the 1970s, Walter Gilbert and Frederick Sanger developed methods to sequence DNA and RNA molecules, which provided researchers with a powerful tool to study gene expression at unprecedented resolution.

In 1977, Phillip Sharp and Richard J. Roberts independently discovered 'split genes': large sections of genes were transcribed together before being chopped up into much smaller pieces called exons.

In the 1980s, the roles of small RNAs began to be revealed. These molecules were found to be involved in many cellular processes including gene silencing and post-transcriptional regulation of mRNA stability. The discovery of microRNAs (miRNAs) further highlighted their importance in regulating gene expression with an estimated 30% of human genes being regulated by miRNA at any one time.

Since then, RNA has been studied intensively as a powerful regulator of gene expression, playing key roles in cell development and differentiation. This has led to tremendous advances in the understanding of how genetic information is processed and transmitted from generation to generation. As our knowledge continues to grow, so too do the potential applications for this powerful molecule.

Origins Of RNA Interference

RNA interference (RNAi) is a biological process in which small pieces of double-stranded RNA (dsRNA) are used to regulate the expression of specific genes. This natural cellular process was first described in 1998 by scientists Andrew Fire and Craig Mello, who subsequently shared the 2006 Nobel Prize in Physiology or Medicine for their work on RNAi.

The discovery of RNAi has enabled scientists to better understand how certain genes are regulated and to create new treatments for many diseases. The mechanism underlying this phenomenon involves the binding of small dsRNAs to specific sequences of messenger RNAs that code for proteins, resulting in either cleavage or translational inhibition. This allows researchers to deliberately silence target gene expression within an organism without affecting any other genes. This precise and specific control of gene expression has made RNAi an invaluable tool in the study of gene function and biology. Additionally, it is being explored as a potential therapeutic approach to treat many diseases caused by abnormal protein production or accumulation.

RNAi is used widely throughout research laboratories across the world, allowing for greater insight into gene regulation and disease etiology. Its use also holds great promise for treating certain human diseases that are caused by overexpression of a single gene or set of genes. Through further understanding of this natural cellular process, scientists may be able to develop more effective treatments for many debilitating conditions.

The Structure And Function Of RNA

RNA, which stands for Ribonucleic Acid, is an essential component of the cell's genetic machinery. It helps to form proteins and plays a role in many other biological processes. The structure of RNA is composed of three parts: a 5' phosphate group, a sugar-phosphate backbone, and a 3' hydroxyl group.

The primary function of RNA is to carry genetic information from DNA and translate it into proteins with the help of ribosomes. This process occurs through transcription and translation. During transcription, the sequence of bases on one strand of DNA is copied onto an mRNA molecule (which contains uracil instead of thymine like DNA). Through translation, this message is decoded by ribosomes and translated into a protein.

The properties of RNA are quite different from those of DNA. For example, RNA is single-stranded while DNA is double-stranded; this makes it more flexible and reactive than DNA. Additionally, RNA molecules tend to be much smaller than their DNA counterparts, which allows them to move around easily within the cell. Furthermore, RNA is able to form secondary structures such as hairpins and stems that help stabilize its shape and function. Lastly, some RNAs have catalytic activity - they can bind to specific substrates and carry out biochemical reactions.

Overall, the structure, function, and properties of RNA all work together in order for it to perform its vital role in the cell's genetic machinery. Its flexibility and reactivity makes it essential for transcription and translation, while its ability to form secondary structures allows it to stabilize its shape and function. Additionally, some RNAs have catalytic activity which gives them the ability to bind to specific substrates and carry out biochemical reactions.

These characteristics make RNA an essential component of life as we know it.

The structure, function, and properties of RNA are essential for its role in the cell's genetic machinery. Its flexibility and reactivity make it possible for transcription and translation to occur, while its ability to form secondary structures helps to stabilize its shape and function. Furthermore, some RNAs have catalytic activity which allows them to bind to specific substrates and carry out biochemical reactions. All these characteristics work together in order for RNA to play an important role in life as we know it. Without its structure, function, and properties RNA would be unable to perform the vital tasks that keep all living things alive.

The chemical nature of RNA.

RNA is an organic molecule composed of ribonucleotides. It serves as a genetic material in some viruses and its structure is different from DNA, containing only a single strand instead of two strands found in the latter. RNA molecules have four main components: ribose sugar, phosphate groups, nitrogenous bases and functional groups. The sugar component makes up the backbone of the molecule while the nitrogenous bases are attached to it. The phosphate groups provide stability to the backbone while functional groups interact with other molecules for various activities such as enzyme catalysis or protein synthesis. In addition, each type of RNA has its own unique sequence that determines its function and structure. For example, messenger RNA (mRNA) contains instructions to make proteins while transfer

RNAs (tRNAs) carry amino acids to build proteins. Thus, understanding the chemical nature of RNA is essential in order to perceive its functions and diverse roles in biology.

The folding of RNA molecules.

The folding of RNA molecules is a complex process that helps the ribonucleic acid (RNA) to assume its specific three-dimensional structure. This is an important step in gene expression, as it is essential for the correct functioning of proteins and other cellular processes. During this folding process, various interactions between different regions of the RNA molecule take place, like hydrogen bonding and electrostatic forces. The result of this intricate process is a compact and highly structured molecule with distinct properties such as stability and catalytic activity. A better understanding of how RNA folds can provide insights into how cells regulate gene expression and help scientists design new medicines or therapies for diseases caused by malfunctioning proteins.

In order to study the folding process, researchers have developed several techniques, including X-ray crystallography and nuclear magnetic resonance (NMR). These allow them to map the three-dimensional structure of the RNA molecule and analyze its interactions. For example, X-ray crystallography has been used to determine the structure of ribosomes, which are essential for protein synthesis. NMR can also be used to study details like hydrogen bonding between certain regions of an RNA molecule.

Overall, studying the folding of RNA molecules is a complex and important task in biology that helps us understand how cells function and design new therapies for diseases. With advances in technology, researchers will likely develop better tools for analyzing this process. This could lead to more effective treatments for diseases caused by malfunctioning proteins.

Synthesis and Stability of RNA.

Synthesis and stability of RNA are two very important concepts in molecular biology. Synthesis of RNA involves the production of molecules from nucleotides that form the building blocks of the genetic information contained within every cell. Once produced, these molecules must remain stable to ensure proper functioning of the cells. This stability is essential for both normal cellular processes as well as replication and storage of genetic information over generations.

The process of synthesizing RNA starts with transcription, during which a molecule of DNA is read by an enzyme called RNA polymerase and copied into a molecule called messenger RNA (mRNA). This mRNA carries copies of instructions, encoded in the DNA sequence, to ribosomes where it will be translated into proteins. After translation, the RNA molecule is then processed and stabilized by enzymes such as ribonucleases, which help fold it into its more stable double-stranded form.

Stability of RNA molecules is also important for ensuring proper functioning of the cells. Without stability, genetic information contained in the DNA would be quickly lost or degraded. To maintain stability, many types of proteins interact with the RNA molecule to ensure that it remains intact and correctly folded. These stabilization proteins protect against degradation caused by UV radiation, temperature changes, and other environmental factors that can cause damage to an unstable molecule.

Synthesis and stability of RNAs are essential components of molecular biology and understanding them helps us gain insight into how cellular processes work.

2.04

Mitosis and Miosis

Mitosis and meiosis are two types of cell division that result in the formation of new cells. Mitosis is a process by which a single parent cell divides into two daughter cells that have the same number of chromosomes as the original cell. This type of cell division occurs during growth, repair and reproduction in organisms with more complex cellular structures. In contrast, meiosis involves the production of four haploid (having half the normal chromosome number) daughter cells from one diploid (having twice the number of chromosomes) mother cell. Unlike mitosis, meiosis only occurs in certain types of sexual reproduction and is used for creating genetic diversity among offspring.

During mitosis, one parent diploid cell first duplicates its genetic material, so that each of the daughter cells will receive a full set of chromosomes from both parents. The parent cell then divides into two new daughter cells through a process called cytokinesis, whereby the cell membrane is pinched and divided to form two new cells. Both of these new daughter cells are genetically identical to the original parent cell they derived from.

Meiosis on the other hand involves two rounds of cell division with only one round of duplicate genetic material – meaning that each daughter cell receives a random selection of genes from both parents. This results in four haploid cells with half the number of chromosomes as the original mother cell. The first round of meiosis (meiosis I) separates homologous pairs of chromosomes so that each daughter cell has one of each pair. The second round (meiosis II) further separates the chromosome pairs, with each daughter cell receiving a unique combination from both parents. This type of genetic diversity helps to ensure that offspring have varied and advantageous traits for survival in different environments.

Overall, mitosis and meiosis are both important processes for cellular reproduction, but they serve two very different functions. Mitosis is used for growth and repair, while meiosis produces genetically diverse offspring which is critical for sexual reproduction.

History

Mitosis and meiosis are two types of cell division processes that occur in all living organisms. Mitosis is a process by which a single parent cell divides into two daughter cells, each having the same genetic information as the parent cell. Meiosis, on the other hand, is a type of cell division in which a single parent cell splits into four daughter cells with half the number of chromosomes as the parent cell.

The history of mitosis and meiosis dates back to 1837 when Theodor Schwann first described these processes under a microscope during his work on animal tissue. About 40 years later, Belgian biologist Edouard van Beneden discovered how chromosomes were moved from one generation to another through meiotic divisions.

He showed that each chromosome of the parent cell divides into two new chromosomes, called chromatids, which are passed on to the daughter cells.

The first detailed description of mitosis was given by German biologist Walther Flemming in 1882. He named it "mitosis" and described how the nucleus divides into two parts during this process. Flemming also explained how chromosomes line up along the equator of a spindle-like structure before being moved to opposite poles.

It took more than 50 years for scientists to understand the full details of meiosis and its mechanism. The main breakthrough came in 1956 when J.M. Griesemer and A.J. Levan proposed their theory about the structure of chromosomes. They proposed that each chromosome is composed of two identical parts, which they called "sister chromatids". This theory was later confirmed by microscopy and genetics experiments.

Today, scientists have a better understanding of mitosis and meiosis processes thanks to various advances in microscopy and genetics. However, there is still more research to be done in order to fully understand these complex cell division processes.

This knowledge can help us gain insights into how cells divide and develop during growth as well as how diseases like cancer manifest themselves at the cellular level.

Occurrence in Eukaryotic Life Cycles.

Mitosis and meiosis are involved in different stages of the eukaryotic life cycle. Mitosis is the process of cell division that occurs during growth and development, allowing for cells to replicate and produce new tissues. Meiosis on the other hand occurs during sexual reproduction and results in the formation of gametes or sex cells. These gametes contain half the number of chromosomes as compared to somatic cells, which ensures genetic variation when two gametes from different individuals combine during fertilization.

Mitosis also plays a role during regeneration, after injury or disease where it allows for the regrowth of lost or damaged tissues whereas meiosis has no role in this process. Furthermore, mitosis is used in cloning techniques when an exact genetic replica of a single organism is required.

Overall, both mitosis and meiosis are essential processes that enable the continuation of life on Earth by allowing for growth, development and variation in living organisms. Without them, evolution would not be possible.

Phases.

Mitosis is a process by which a single cell divides into two identical daughter cells. It is the first stage of cellular division, followed by cytokinesis. The steps of mitosis are interphase, prophase, metaphase, anaphase and telophase.

Interphase: This is the longest phase of the cell cycle in which DNA replication occurs. During this phase, the nucleus contains two copies of every chromosome that are connected at centromeres.

Prophase: At this point in mitosis, the nucleus begins to break down and spindle fibers begin to form between the two copies of each chromosome. Centrioles migrate towards opposite ends of the cell and nucleoli become less visible as chromatin is condensed to form visible chromosomes.

Metaphase: During this stage, the chromosomes line up in the center of the cell and align at a point called the metaphase plate. The spindle fibers attach to each chromosome and help guide them into position.

Anaphase: In anaphase, the centromeres of each chromosome split apart and begin to move towards opposite ends of the cell. This movement is facilitated by motor proteins and is driven by ATP energy from the mitochondria.

Telophase: At this point, two new nuclei have formed as a result of cytokinesis. The nuclear envelopes are re-formed around each nucleus and chromatin returns to its uncoiled state. The cell is now ready to exit mitosis and enter the next stage of the cell cycle.

Miosis is a form of nuclear division that produces four haploid (n) daughter cells from one diploid (2n) parent cell as part of sexual reproduction in eukaryotic organisms. It includes two successive divisions, meiosis I and meiosis II, each consisting of five distinct stages: prophase I, metaphase I, anaphase I, telophase I and cytokinesis, followed by the same stages in meiosis II.

Prophase I: During this stage, homologous chromosomes (a pair of chromosomes with similar genetic makeup) begin to move together and exchange sections via a process known as crossing over. Chromosomes also condense and become visible under a microscope.

Metaphase I: This stage is marked by the alignment of homologous chromosomes along the metaphase plate. Spindle fibers from opposite poles attach to each pair of chromosomes, helping guide them into position.

Anaphase I: The centromeres of each pair of homologous chromosomes split apart and move towards opposite poles, driven by motor proteins powered by ATP energy from mitochondria.

Telophase I & Cytokinesis: At this point, two new daughter cells are formed as a result of cytokinesis. The nuclear envelopes reform around each nucleus and chromatin returns to its uncoiled state.

Meiosis II: This is the second stage of miosis and consists of the same five steps as meiosis I, but without the exchange of genetic material via crossing over. The two daughter cells formed in meiosis I then divide into four new haploid cells.

At the end of miosis, there are four new haploid daughter cells that each contain half the number of chromosomes present in the original parent cell. These genetically unique cells can then go on to form gametes for sexual reproduction.

The process of mitosis and meiosis is an essential part of the cell cycle, allowing for growth and development in multicellular organisms. Without these processes, organisms would not be able to reproduce or develop properly. By understanding the steps involved in both mitosis and meiosis, scientists are better able to study cellular division and understand how it contributes to the life cycle of different organisms.

PART-03

3.01

New world

The New World has ushered in a new era of human existence. Compared to the Old Era, the dominant way of life today is largely based on technology and modern conveniences. The industrial revolution has led to mass production, while advances in communication and transportation have streamlined global trade networks. This combination has enabled us to access an unprecedented level of luxury goods, consumer products, and services.

In addition, the rise of digital technologies has completely transformed the way we work, learn, communicate, shop, and entertain ourselves. We can now connect with almost anyone from any corner of the world with just a few clicks or taps on our phones or computers. Social media platforms such as Facebook and Twitter allow us to instantly share our thoughts and experiences with millions of people.

The New World has also brought about an increased awareness of global socio-economic issues and a heightened sense of responsibility for the state of our planet's resources and ecosystems. This newfound consciousness has led to more sustainable practices in many industries, from energy and food production to fashion design. We are beginning to recognize our need to conserve natural resources for future generations, while still enjoying the abundant comforts that modern technology can provide us.

To summarize, the New World is characterized by rapid technological advancement, increased globalization, and a heightened sense of environmental responsibility. We now have access to remarkable levels of comfort and convenience, all while striving towards creating a better world for ourselves and our descendants.

Advanced time and world.

Advanced Time and World is a revolutionary concept that combines the idea of virtual and physical worlds to create a new era of connection. By connecting the two environments—virtual and physical—users can explore a variety of activities, such as online gaming, immersive storytelling, personal communication, education, collaboration, and more. The advanced time-based features allow users to experience events from multiple perspectives in real-time or delayed motion for added immersion. Additionally, this technology provides an opportunity for businesses to engage customers with interactive experiences through digital media campaigns. With its vast potential use cases, Advanced Time and World is revolutionizing how we interact with each other and our environment.

This technology is not only beneficial for businesses, but also for individuals. It helps them gain an understanding of the events and process that define their lives. By connecting people with each other in a new way, Advanced Time and World provides users a chance to better understand their environment. This gives them access to valuable insights from multiple perspectives, which can be used to create meaningful connections to help them achieve personal goals.

The possibilities of Advanced Time and World are almost limitless - from creative projects such as animation and film production to research initiatives like crowd-sourced data collection. Its potential use cases span across industries, allowing companies to take advantage of the innovative platform for marketing purposes or educational experiences. With its ability to connect physical and digital worlds, Advanced Time and World is revolutionizing the way we interact with our environment.

By leveraging this technology to its fullest potential, businesses can bridge the gap between physical and virtual worlds and create meaningful experiences for their customers. With its vast potential applications, Advanced Time and World has already begun to reshape how we engage with each other on a digital level. As more users explore this new realm of technology, there are no limits to what it can achieve in connecting us all.

Thanks to its innovative features and powerful capabilities, Advanced Time and World is becoming increasingly popular among people around the world. It provides an unprecedented opportunity for individuals to connect with one another in a new way while also allowing businesses to capitalize on its potential use cases. As this technology continues to evolve, it is sure to revolutionize the way we interact with our environment and each other.

The possibilities of Advanced Time and World are endless - from creative projects to research initiatives, its potential use cases span across industries. With its ability to connect physical and digital worlds, it provides a unique opportunity for businesses to bridge the gap between the two while also creating immersive experiences for their customers. We can only imagine what else Advanced Time and World will be able to achieve in connecting us all as it continues to grow in popularity.

Utilizing this revolutionary technology is an essential step towards progress – no matter how small or large your business may be. By leveraging the powerful capabilities of the technology, you can provide your customers with a unique experience that is sure to make an impact. The possibilities of Advanced Time and World are limitless - with its ability to connect physical and digital worlds, it has already begun to revolutionize how we interact with our environment. As this technology continues to evolve, its potential will only continue to grow.

It's no surprise that Advanced Time and World is becoming increasingly popular among people around the world. With its innovative features and powerful capabilities, businesses have access to an unprecedented platform for marketing purposes or educational experiences – creating meaningful connections between physical and digital worlds in order to achieve success. Now more than ever, it is essential for businesses to understand what this revolutionary technology can do for them and take the necessary steps to leverage its potential use cases. By doing so, they can ensure that their business thrives while also providing customers with an unparalleled experience.

Advanced Time and World is revolutionizing how we interact with our environment – providing us with a unique opportunity to better understand and engage with each other. With its ability to connect physical and digital worlds, it has already begun to reshape how we use technology in order to create meaningful connections and experiences. As this technology continues to evolve, the possibilities are endless – giving businesses an unprecedented platform for success and connecting us all on a deeper level than ever before. It's no wonder why Advanced Time and World is becoming increasingly popular among people around the world!

Problems and issues of advanced world.

Advanced world faces various problems and issues due to the rapid development of technology and its utilization in everyday lives. One of the most pressing issues is environmental degradation, due to overuse of resources and pollution. Climate change is an increasingly serious issue that has been linked to global warming, sea level rise, desertification, air quality decline, and biodiversity loss. Moreover, technological advances have created economic disparities in many countries as access to digital resources increases inequality between different socioeconomic groups. Another concern is the potential for artificial intelligence (AI) technologies to displace human labor or be used for unethical purposes. Additionally, cyberattacks are a growing threat as companies become more reliant on data-driven processes and rely heavily on computers in their operations. Finally, increasing digitalization of everyday life has created privacy and security issues as individuals share sensitive data online. These issues must be addressed in order to ensure that the advanced world is able to progress sustainably and ethically.

Overall, the problems and issues of the advanced world are complex, but not insurmountable. Governments, organizations, and individuals must work together to address these challenges and create systems for a better future. By investing in renewable energy sources, improving digital literacy initiatives, protecting consumers from cyberattacks, and creating regulations on AI usage, we can all make a difference in addressing the problems of the advanced world. With proper planning, collaboration, and implementation of everyday life has led to an increase in privacy concerns, as data breaches become more common and personal information is not always secure. As the world continues to progress technologically, addressing these problems and issues will be essential for achieving a sustainable future.

Overall, advanced world presents several complex problems and issues that need to be tackled with careful consideration. Without proper solutions, it is likely that environmental degradation, economic disparities, AI misuse, cyberattacks, and privacy violations could worsen over time. It is crucial that we continue to work together towards developing meaningful solutions that can promote a better future for all.

When it comes to the benefits of advanced technology, improved healthcare systems are often cited as one of the main advantages. Technological advances have led to improvements in diagnosis and treatments, allowing more people access to healthcare services. Additionally, digital innovations are making it easier for medical professionals to collaborate on research projects and share data quickly. This is helping advance medical knowledge and driving new discoveries in the field of medicine. Furthermore, increased automation is leading to enhanced efficiency across many industries, reducing costs and increasing profits for businesses. Finally, advanced technology allows us to stay connected with one another from anywhere in the world through digital platforms like social media and video conferencing. These advantages must be balanced against the potential risks that come with technological advancements if we want to create a better future for all.

Ultimately, the advanced world presents an array of opportunities and challenges that must be addressed in order for us to create a better future. By investing in renewable energy sources, expanding digital literacy initiatives, protecting consumers from cyberattacks, and creating regulations on AI usage, we can all work together towards a more sustainable and equitable future. With proper planning and collaboration between governments and stakeholders, we can ensure that the advanced world continues to progress sustainably and ethically.

3.02

Modern life

Modern life style of humans nowadays is characterized by a number of factors. These include increased urbanization, in which more and more people are living in cities; an ever-increasing pace of work and the availability of technology to stay connected even during leisure time; reliance on processed food, with convenience often being placed over health considerations; increasing levels of pollution, especially air pollution, due to the burning of fossil fuels for energy; less physical activity due to sedentary lifestyles; a greater tendency towards consumerism and materialism as opposed to spiritual values; and shorter attention spans due to constant bombardment from screens and social media. All these elements combine together to create modern life style that can be both beneficial as well as detrimental depending on how they are used. It is up to each individual to make the right decisions when it comes to incorporating a modern lifestyle into their daily lives.

Meanwhile, there are benefits associated with modern life style as well such as increased access to information and education, wider availability of goods and services, better healthcare facilities and more opportunities for leisure activities. All these can add significant value to our lives if incorporated in a balanced way. Thus, while modern life style needs to be used judiciously, its advantages should not be overlooked either.

Overall, modern life style is here to stay; however, it is important that we use it wisely and create healthy balance between convenience and wellbeing in order to lead meaningful lives that bring us joy and satisfaction.

In conclusion, modern life style has become a reality of our times. Even though it can be challenging to manage and requires careful thought and balance in order to make the right decisions, its advantages are undeniable. There is immense potential that comes with modern life style, provided that we use it sensibly and for the benefit of our health and wellbeing. By learning how to use these changes to our advantage, we can really reap the rewards and enjoy all of what life in the 21st century has to offer!

The key takeaway here is that although modern life style can be challenging at times, there are many opportunities associated with embracing such lifestyle if done so responsibly. With the right knowledge and proper management, modern lifestyle can be a great asset in achieving a healthy and meaningful life.

Modern life civilization.

Modern life civilization is defined by its wide variety of features, which encompass everything from technological advances and the rise of cities to increased access to education and healthcare. In modern times, people have more opportunities than ever before—not only to make a living through their chosen field or profession but also to explore new cultures, ideas, and concepts. Technology has enabled us to communicate faster than ever before; we now have access to information that was previously unavailable; transportation options allow us more freedom of movement; and our economies are intertwined in ways never seen in history. Furthermore, globalization has led to an increase in diversity across nations as well as within many societies. This phenomenon has enriched our lives in countless ways, from advancing our knowledge about distant cultures to broadening our understanding of the world we inhabit. In short, modern life civilization is characterized by its multiple advances and possibilities.

These opportunities have had huge impacts on our lives, from changes in the way we work to advancements in healthcare and medicine. Education has become more accessible than ever before, allowing us to learn new skills or gain qualifications which are essential to advancing careers; this also means that more people are entering university and college than ever before. Technology has enabled us to stay connected no matter where we are, making remote working a reality for millions of people around the world. Sustainable practices have also been integral to modern life civilization, as individuals, organizations, governments, and corporations take responsibility for reducing their environmental impact and protecting resources for future generations.

In conclusion, modern life civilization is defined by its wide range of advances and features which have allowed us to explore the world like never before. We're living in an era where technology has enabled us to stay connected with each other, access resources from all around the globe, and make a positive impact on our environment. All this indicates that the future looks bright as we continue to innovate and develop new solutions for the problems of today's modern world.

Cultures and religions.

Modern life culture and religion can be defined as a set of beliefs, values, practices, customs, and norms shared by a group of people. These cultures and religions influence how individuals view the world around them, how they interact with others, and what they believe is right and wrong. Modern life cultures are also shaped by societal institutions such as schools, governments, businesses, media outlets, churches or other places of worship. Religion typically encompasses spiritual beliefs in gods or deities as well as rituals that involve prayer, fasting or meditation.

Different cultures have different sets of values and belief systems which shape their views on education, family dynamics, gender roles and economic structures. For instance Confucianism is centered around reverence for elders and ancestors while Hinduism is more focused on the individual's relationship with divine energy. Secular cultures which are not tied to any one religion, focus heavily on materialistic pursuits and elevating self-interest over collective well-being.

In today's world, many cultures and religions are becoming increasingly intertwined due to globalization, technology advancements and migration. As a result, new cultural norms and traditions are formed as different societies interact with each other. This process of blending cultures often leads to hybrid practices that incorporate elements from both sides while discarding aspects that no longer serve them. Ultimately, modern life culture and religion shape how we interact with each other as a society. By understanding the nuances of various cultures and religions around us, individuals can better appreciate the diversity of the world we live in.

Modern life culture and religion provide a source of stability, comfort, and support to people from all walks of life. By following cultural norms and religious practices, individuals can feel connected to their past and present while also having an optimistic view towards the future. For some, cultural and religious expression is also a form of self-expression, allowing them to express themselves in ways that may not be accepted by mainstream society. By exploring different cultures and religions, we can gain insight into how societies have evolved over time in relation to certain values or belief systems. Additionally, understanding different cultures can help us gain a better appreciation of our own culture and beliefs as well as provide empathy towards those from different backgrounds.

Ultimately, modern life cultures and religions continue to shape the world we live in today. By exploring these cultures and religions, we can gain greater insight into how the world functions which can help us better understand ourselves and others around us. Through this process, individuals can come to appreciate the beauty of diversity that exists in also aid in developing greater empathy towards others and can help build bridges between different communities.

Overall, modern life culture and religion are ever-evolving concepts that shape our identity as individuals and various societies around the world. By understanding these cultural and religious aspects of our lives, we can create a more harmonious society for everyone.

3.03

Health issues

Health is an essential part of life. It refers to the well-being of our physical, mental, and social wellbeing. However, there are a variety of health issues that can affect individuals or entire societies. These issues may include chronic diseases such as cancer and diabetes, infectious diseases such as HIV/AIDS, environmental hazards like air pollution or water contamination, occupational hazards like unsafe working conditions, lack of access to quality healthcare services, poverty and inequality within populations, unhealthy lifestyles which can lead to obesity and many other illnesses, self-harm behaviors including drug addiction and alcoholism; etc. All these health issues have serious consequences that can affect not only the individual's quality of life but also the overall economic productivity in a society. Therefore it is important to recognize and address these health issues in order to ensure everyone can enjoy a healthy and prosperous life.

Poorly lifestyles which can lead to obesity and many other illnesses, self-harm behaviors including drug addiction and alcoholism; etc. All these health issues have serious consequences that can affect not only the individual's quality of life but also the overall economic productivity in a society. Therefore it is important to recognize and address these health issues in order to ensure everyone can enjoy a healthy and prosperous life.

Physical health.

the risk of developing heart disease or stroke. Additionally, it can reduce depression and anxiety levels, improve cognitive function, reduce inflammation levels in the body, enhance immunity, and help maintain healthy cholesterol levels.

Good nutrition is another key element of physical health. Eating a balanced diet helps provide the body with essential nutrients and energy for daily functioning. The recommended dietary guidelines recommend eating fruits, vegetables, lean proteins, whole grains, low-fat dairy products, and healthy fats in moderate amounts. It is also important to limit processed foods, added sugars, refined grains, and saturated fat to maintain physical health.

Getting enough sleep is an important part of staying physically healthy as well. Sleep allows the body to recharge and restore itself after a day of activity. Adults should aim for 7-9 hours of quality sleep per night while children need even more than that (as much as 10-12 hours for school-aged children). Not getting enough sleep can lead to fatigue, impaired immunity, slower reaction time, poor decision-making, and a decreased ability to concentrate.

Maintaining proper hygiene is another important component of physical health. Regular showering or bathing helps remove dirt and bacteria from the skin while brushing teeth twice daily helps fight plaque buildup and prevent tooth decay. Additionally, it is important to wash hands frequently throughout the day to help reduce the spread of germs and illnesses.

Finally, stress management plays a key role in overall physical health. Engaging in regular relaxation activities like yoga, meditation, or deep breathing can help reduce stress levels and improve mental clarity. Additionally, getting plenty of exercise and eating a balanced diet can also help combat stress as they both release endorphins which are mood boosters that help to relax the body.

By following simple recommendations for physical health, such as getting enough sleep, exercising regularly, eating a balanced diet, and managing stress levels, we can improve our overall health and well-being. With these guidelines in place, we can keep our bodies functioning optimally and experience improved quality of life overall.

Mental health.

Mental health is a state of emotional and psychological well-being in which an individual is able to cope with the ordinary stresses of life, work productively and fruitfully, and make meaningful contributions to their community. Mental health includes our emotional, psychological, and social wellbeing; it affects how we think, feel, and act. It also helps determine how we handle stress, relate to others, and make choices. People with good mental health are able to fulfill personal potentials; successfully deal with life's challenges; form healthy relationships; cope with adversity; help others in need; live responsibly; experience enjoyment and satisfaction from life experiences. Poor mental health can prevent people from functioning properly in everyday situations.

Common indicators of poor mental health include sleep disturbances, difficulty concentrating, mood swings, feelings of loneliness or depression, loss of interest in activities once enjoyed, and thoughts of suicide. Mental health disorders such as anxiety and depression can have a devastating impact on an individual's life if left untreated. Treatment options such as counseling and medication can help improve mental health and reduce the symptoms of mental illness. It is important to recognize signs of poor mental health in order to seek treatment before it becomes more serious. Taking steps towards improving one's mental health is key to achieving overall wellbeing. With the right support, individuals with poor mental health can work towards leading a healthier life. It is never too late to start taking steps towards recovery from poor mental health - so don't wait any longer! Reach out for help if you or someone you know is struggling. There are always people willing to offer support and guidance, and it could make all the difference in your life.

Mental health is a complex concept that encompasses many different facets of our lives. It includes emotional, psychological and social wellbeing; how we think, feel and act in response to daily stressors; our ability to manage our emotions and relationships, form healthy coping skills, cope with adversity, take responsibility for ourselves and others, experience pleasure in activities we engage in; as well as how we interact with our environment. Mental health can be seen as an umbrella term that covers a wide variety of individual experiences from anxiety to depression to substance abuse.

Mental illnesses are medical conditions that affect a person's thinking, feeling, moods and behavior.

Mental health disorders can range from mild to severe and can have a significant impact on an individual's overall wellbeing. Symptoms of mental illness include changes in mood or behavior, feelings of sadness or emptiness, difficulty sleeping or concentrating, disruptive thoughts or behaviors, and substance abuse. It is important to recognize the signs of poor mental health so that individuals can seek treatment before it becomes more serious. Treatment options such as psychotherapy, medication and lifestyle modifications can all be effective in helping improve one's mental health. Taking steps towards recovery is key to achieving a healthy life balance - so don't wait any longer! Reach out for help if you or someone you know is struggling. With the right support, individuals with poor mental health can work towards leading a healthier life. Remember, it is never too late to start taking steps towards improving one's mental health - so don't wait any longer! There are always people willing to offer support and guidance, and it could make all the difference in your life.

Mental health is an important part of overall wellbeing, and it is essential to recognize its importance. In order for us to live happy and fulfilling lives, we must all strive for better mental health by seeking help when necessary, maintaining healthy relationships with friends and family members, engaging in activities that give us joy and fulfillment, and learning how to manage stress effectively. By prioritizing our mental health, we can ensure that our overall wellbeing is strong and resilient.

Social health.

Social health is an important part of overall wellbeing and can be defined as the ability to maintain positive relationships with others, have meaningful connections with family and friends, participate in communities or activities that support one's sense of purpose, and take responsibility for one's physical and emotional health. It encompasses all aspects of how people interact within their environment—including both personal relationships and the larger social structures they inhabit. Social health requires a person to understand their own needs, limits, values, beliefs, emotions, strengths, weaknesses, and limitations. Along with this awareness comes the capacity for self-care—the ability to regulate emotions responsibly (not bottling them up or letting them run wild), express oneself in healthy ways without lashing out at others verbally or physically, and cope with stress in healthy ways (exercise, journaling, talking to a therapist or trusted friend). Ultimately, social health is about how one uses communication skills to foster meaningful relationships and develop strong connections with those around them. When someone has good social health they are able to form relationships that are not only meaningful but also beneficial—for both themselves and the people with whom they interact. In addition, having good social health means being self-aware enough to recognize when it's time for self-care or when extra help from a mental health professional may be needed.

In conclusion, social health is an important part of overall wellbeing that involves the ability to interact positively with others on a personal level as well as within larger social structures, and practice self-care when needed. With a good understanding of one's needs, limits, values and emotions, and the ability to use communication effectively, an individual with strong social health can create meaningful relationships that benefit both themselves and those around them.

Social health disorders are a type of mental health disorder that is characterized by difficulties in an individual's ability to interact socially with others. This can include struggling to create and maintain healthy relationships, difficulty understanding basic social cues, lack of empathy and/or sympathy towards others, as well as feeling uncomfortable or awkward in social situations. People with social health disorders can also experience symptoms such as anxiety and depression due to their struggles with forming-and maintaining- relationships.

In order to diagnose someone with a social health disorder, it is important for doctors and mental health professionals to assess the patient's history and behavior over a period of time. They may ask about any past or present stressors that could be contributing to the individual's feelings, as well as their current coping strategies. Treatment for social health disorders typically consists of therapy and/or medication to help the individual manage their symptoms. With the help of a mental health professional, individuals struggling with social health disorders can learn healthy coping skills, communication techniques, and gain insight into their behavior in order to improve their ability to function in social settings.

Overall, the goal of treating social health disorders is to increase emotional awareness and self-regulation in order to create an environment that promotes positive relationships within one's life. While it can be a difficult journey, with proper treatment and support anyone living with a social health disorder can lead a happier and more fulfilling life.

Viral, Bacterial, and Infectious diseases.

Viral diseases are caused by viruses, which are microscopic organisms made up of genetic material and a protein coat. They invade living cells in the body and use their hosts' cellular machinery to replicate themselves. Common viral diseases include the common cold, influenza, chickenpox, herpes, HIV/AIDS, SARS and MERS-CoV.

Bacterial diseases are caused by bacteria - single-celled microorganisms that can exist either as independent (free-living) organisms or as parasites (dependent on another organism for life). These microorganisms are found everywhere in nature and can cause many types of infections such as strep throat, tuberculosis and urinary tract infections.

Infectious (contagious) diseases are those caused by organisms, such as bacteria, viruses and fungi, that can be spread directly or indirectly from one person to another. Examples of infectious diseases include measles, mumps, rubella, hepatitis A and B, and influenza. These illnesses often cause fever and other symptoms such as coughing, sneezing or rashes.

By understanding how these diseases are caused and spread it is possible to reduce the risk of infection. People can prevent many viral infections through immunization (vaccination). Good hygiene practices such as hand-washing with soap and water are also important for preventing the spread of illnesses. Additionally, antibiotics can help treat some bacterial infections but have no effect on viral infections since they cannot attack the virus itself. Proper diagnosis is also essential for treating infectious diseases effectively. Taking all these prevention and treatment measures can help reduce the spread of viral, bacterial, and infectious diseases.

Fungal infections are caused by a group of organisms called fungi which includes yeasts and molds. Fungal infections can affect almost any part of the body but are more common in warm, moist areas such as the mouth and genitals. Common fungal infections include athlete's foot, jock itch and yeast infections. Treatments for fungal infections vary depending on the type of infection but may include antifungal medications or creams applied to the affected area or oral medications taken by mouth. Prevention measures such as good hygiene practices can also help reduce the risk of fungal infections.

By understanding the causes and methods of transmission of viral, bacterial, and infectious diseases, it is possible to take preventive measures that can reduce the risk of infection. Vaccinations, good hygiene practices such as hand-washing with soap and water, and proper diagnosis are all important tools in preventing and treating these illnesses. Additionally, appropriate treatments such as antibiotics for bacterial infections or antifungal medications for fungal infections should be taken to reduce the spread of disease. Taking all these steps can help promote better health outcomes in individuals who may be at risk for these illnesses.

Common modern life diseases.

Modern life diseases are often caused by an unhealthy lifestyle, environmental pollutants, and genetic factors. They range from chronic conditions like heart disease, diabetes and cancer to infectious diseases such as influenza, herpes, and HIV/AIDS.

Heart Disease: A common modern life disease is coronary artery disease or heart disease. This is a condition in which plaque builds up on the walls of the arteries that carry blood to the heart muscle. Over time, this buildup causes narrowing of the arteries leading to reduced blood flow and can eventually cause a heart attack if untreated. There are several risk factors for developing this condition including smoking, obesity, poor diet and lack of physical activity. Treatment typically includes lifestyle changes such as eating healthy foods, exercising regularly and quitting smoking. In some cases, medications or surgery may also be needed.

Diabetes: Diabetes is a chronic condition in which the body does not produce enough insulin to control blood sugar levels. This can lead to high blood sugar levels and an increased risk for complications such as, as well as the use of medications to lower cholesterol and blood pressure.

Diabetes: Another modern life disease is diabetes. This chronic condition is caused by high levels of sugar in the blood and affects how the body uses energy from food. It occurs when the pancreas does not produce enough insulin or when the body does not properly respond to its own insulin. Symptoms can include increased thirst, fatigue, frequent urination and blurry vision. Treatment includes regular monitoring of glucose levels, lifestyle modifications such as eating healthy foods and exercising regularly, and medication to control blood sugar levels.

Cancer: Cancer is a group of diseases characterized by uncontrolled cell growth due to mutations in genes that regulate cell division. It can develop anywhere in the body and can spread to other parts of the body. Common types include lung, breast, colon, and prostate cancer. Risk factors for developing certain cancers include smoking, obesity, age and genetics. Treatment depends on the type and stage of cancer but typically includes radiation therapy, chemotherapy or surgery in combination with lifestyle changes such as eating a healthy diet and getting regular exercise.

Influenza: Influenza is a contagious respiratory illness caused by viruses that infect the nose, throat and sometimes lungs. Symptoms can range from mild fever to more severe symptoms such as body aches, chills and fatigue. Vaccines are available each year to help protect against infection from the most common strains of influenza virus circulating at the time. Treatment usually includes rest and plenty of fluids as well as over-the-counter medications to help reduce fever and body aches.

Herpes: Herpes is an infectious disease caused by the herpes simplex virus, which affects the skin and mucous membranes. It can cause both genital and oral outbreaks in which painful blisters or sores appear on the affected area. Treatment includes antiviral medications to reduce symptoms such as itching, burning and pain. Additionally, there are drugs available that may reduce the likelihood of recurrences.

HIV/AIDS: HIV (human immunodeficiency virus) is a virus that attacks the immune system and weakens it so that other infections can take hold more easily. AIDS (acquired immunodeficiency syndrome) is the final stage of HIV infection and is characterized by a significant decrease in the immune system's ability to fight infections. There is no cure for HIV/AIDS, but treatments are available that can help manage symptoms and reduce the risk of transmission. These include antiretroviral medications taken daily, as well as regular testing for other sexually transmitted diseases and counseling to promote safer sexual practices.

These are some of the most common modern life diseases, but many more exist such as asthma, hepatitis C and chronic obstructive pulmonary disease (COPD). While each has its own set of causes and treatments, all require lifestyle modifications to reduce risks and improve overall health outcomes. Additionally, early detection through regular screenings for these conditions can drastically improve prognoses.

3.04

A Healthy Life

Good health is essential to living a healthy life. It enables us to do everything from performing routine daily tasks and activities to having the strength and energy needed for exercise, leisure activities, and general wellbeing. Good health encompasses physical, mental, emotional, spiritual, social, and environmental aspects of our lives.

20 tips to live a healthier life.

1. Get enough sleep - Make sure to get seven or more hours of sleep each night. Studies have shown that not getting enough rest can have negative consequences on your physical and mental health.

2. Exercise regularly - Exercise doesn't need to be complicated, it can just be a walk around the block or some stretching at home. The important thing is to make time for it in your daily routine so you create healthy habits and stay active.

3. Eat a balanced diet - Eating healthy meals with all the essential nutrients helps keep our bodies strong and functioning properly. Aim for food rich in vitamins, minerals, proteins and fibre while limiting processed foods as well as saturated fats and sugars found in sweets, fried foods etc.

4. Drink lots of water - Staying hydrated is key for our bodies to work properly and it helps keep us energised throughout the day. So, make sure to drink plenty of water every day and cut back on sugary drinks like sodas and juices.

5. Reduce stress - Stress can have a negative impact on both physical and mental health so it's important to find ways to manage it. This could be through mindfulness, yoga or any activity that can help you relax while keeping your mind clear.

6. Socialise with family & friends - Spending time with people we care about helps improve our mood, gives us positive energy and makes us feel more supported when facing difficult times in life. So, don't forget to make time for your loved ones and enjoy the company of others.

7. Take time for yourself - We all need some time alone every now and then in order to recharge. That could be a hot bath, reading a book or doing any activity that makes you feel relaxed and happy.

8. Practice self-care - Taking care of yourself is always important as it helps boost your confidence and keeps you feeling good inside out. This could mean getting regular haircuts, treating yourself with something special from time to time or simply pampering yourself with face masks and enjoying moments of relaxation at home.

9. Eat breakfast every morning - Eating breakfast is key to kickstart your metabolism and give yourself the energy needed for the day ahead. A good balanced option is oatmeal or fruits combined with some proteins like eggs or yoghurt.

10. Cut back on sugar - Too much sugar can be bad for our health so try to make healthier choices such as using honey, agave syrup or stevia instead of regular sugar.

11. Quit smoking - Smoking not only affects your lungs but it also increases the risk of many illnesses such as cancer and heart disease. If you want to quit, there are many resources available like smoking cessation programs that can help you on your journey.

12. Avoid alcohol - Drinking alcohol in moderation is ok, but drinking too much can lead to serious health problems. So make sure to keep track of your consumption and limit it if necessary.

13. Clean & declutter - Keeping a clean home helps reduce stress levels and makes us feel more organized and productive. It also reduces dust and allergens which in turn improves our overall health.

14. Get enough sunshine - Sunlight is important for synthesizing vitamin D within our bodies, which helps strengthen bones and teeth while boosting mood. So, don't forget to make time for outdoor activities like walking or cycling and enjoy the benefits of getting some sunlight.

15. Take regular breaks - Breaks are essential for our mental wellbeing as they help clear our minds and recharge our energy levels throughout the day. Whether it's just five minutes to stretch your legs or a whole afternoon to do something fun, taking breaks makes a big difference in how we feel afterwards.

16. Manage screen time - Too much time on screens can be detrimental to both physical and mental health so it's important to find ways to manage it properly. This could be by setting limits on devices and apps or avoiding using them right before bedtime.

17. Get regular check-ups - Regular health screenings can help detect any illnesses early on and keep our bodies healthy in the long term. So don't forget to make time for doctor visits or tests as needed.

18. Create a positive environment - Surrounding ourselves with positive people, activities and things helps us stay motivated and productive while improving our overall wellbeing.

19. Avoid junk food - Fried and processed foods are loaded with saturated fats, added sugars, sodium and unhealthy preservatives which can be detrimental to our health if taken too often. So try to avoid them by cooking more meals at home or choosing healthier snacks when eating out instead.

20. Be kind to yourself - Don't forget to be gentle and compassionate with yourself. Everyone makes mistakes and that's ok. So don't forget to forgive yourself, embrace your flaws and recognize the good things about you.

Exercise and Meditation.

Exercise and meditation are two popular activities that have been proven to improve mental and physical health. Both offer a variety of benefits, including improved concentration, better sleep, decreased stress levels, increased energy, and the ability to manage anxiety. Exercise focuses on physical activity while meditation focuses on the mental practice of relaxation and mindfulness.

When it comes to exercise there are many different types of workouts that can be done such as running, walking, weight lifting, yoga or any other type of physical activity. Regular exercise is important for overall health and wellness as it helps strengthen muscles and bones, increase endurance and reduce stress. It also helps with maintaining healthy body weight which in turn reduces risk for chronic diseases like diabetes or heart disease.

Meditation is a mental exercise that can be practiced at any time and requires only a few minutes of quiet time.

It involves focusing on the present moment and using deep breathing techniques to reduce stress and improve concentration. Practicing meditation regularly can lead to emotional well-being, improved relationships with others, better self-awareness, more insight into life's challenges and enhanced creativity.

Exercise and meditation are both beneficial practices that should be incorporated into one's daily routine for optimal physical and mental health. Regular exercise helps maintain physical fitness while meditation increases mindfulness and reduces stress levels. Both activities can help improve quality of life, leading to greater happiness in the long run!

Healthy diet and food.

Eating a healthy diet and incorporating nutritious food into your meals is essential for good physical and mental health. It can help to reduce the risk of chronic diseases such as heart disease and diabetes, as well as providing nutrients that support overall wellbeing. Eating a varied diet that includes fruits, vegetables, whole grains, lean proteins and healthy fats is important. Avoiding processed foods or unhealthy snacks full of sugar, salt, trans fats or added preservatives can also help you maintain a healthy balance in your diet. Physical activity is also important to ensure that your body gets the exercise it needs and maintains its strength. The combination of eating right and getting regular exercise will lead to improved physical condition, increased energy levels, and better overall mental health. Taking care of your body by eating healthy and exercising regularly is the best way to ensure long-term health and happiness.

Making small changes to your diet can have a big impact on your overall wellbeing. Incorporating more fruits and vegetables into your meals, replacing processed snacks with healthier options such as nuts or seeds, choosing lean proteins over red meats, and opting for whole grain cereals rather than heavily sugared varieties are all simple dietary switches that can improve nutrition while still providing delicious options. Additionally, drinking plenty of water throughout the day helps to keep the body hydrated which has many positive benefits for both physical and mental health. Lastly, there are many ways to increase physical activity levels; from taking a walk around the block to joining an exercise class, the key is to find something that fits your lifestyle and allows you to get regular physical activity.

Overall, a healthy diet and active lifestyle are essential for maintaining both physical and mental wellbeing. With some simple changes, it's easy to make sure that your body is getting all of the nutrition it needs while also allowing you to enjoy delicious meals. By incorporating fruits, vegetables, lean proteins, whole grains and other nutritious foods into your diet as well as engaging in regular physical activity, you can ensure that you remain happy and healthy for many years to come.

A healthy meal should consist of a variety of nutrient-rich foods. It should include fruits, vegetables, whole grains, proteins (like lean meats, fish, eggs and legumes), and dairy foods like milk or yogurt. Additionally, it is important to limit processed foods high in sodium, saturated fat and added sugar.

When creating a healthy meal plan for yourself or your family, it's important to look for recipes that are not only nutritious but also creative and delicious. Step-by-step recipes make the process easier so you can create healthier meals quickly. Also consider adding fun variations or twists to classic dishes to add more flavor with fewer calories.

In order to ensure proper nutrition throughout the day, it's important to eat three meals each day and snack when necessary. This schedule helps the body maintain energy levels by providing consistent fuel. Eating regularly also keeps your metabolism functioning at its best and ensures that you are getting all the essential vitamins, minerals and other nutrients your body needs.

Finally, it is important to remember that healthy eating is a lifestyle choice, not a short-term solution or fad diet. Taking small steps towards creating healthier meals and snacks for yourself can add up to big changes over time. With creative recipes, regular eating schedules and mindful decision-making about food choices, anyone can create a delicious and nutritious meal plan for themselves or their family.

Drinking water.

Drinking water is essential to sustaining life. Not only does it replenish our bodies, but it helps to flush out toxins and waste, while providing important electrolytes and minerals that aid in digestion and metabolism. It's also important for maintaining a healthy weight, as drinking water can curb your appetite and make you feel fuller faster. Drinking plenty of water can also help reduce headaches, improve concentration, prevent fatigue, boost energy levels and even improve skin hydration. It's no wonder why the Centers for Disease Control and Prevention recommends drinking eight 8-ounce glasses of water each day!

However, when it comes to health benefits of drinking water, there are some things to keep in mind. While plain water is the best form of hydration, other beverages such as sports drinks and juices can contain electrolytes and minerals that help to replenish lost fluids after periods of heavy exercise. Also, caffeinated beverages can be dehydrating and should be consumed in moderation. Additionally, if you live in an area with poor water quality, investing in a water filter or buying filtered water can help to ensure that the water you're drinking is safe.

Overall, it is clear that drinking plenty of water has numerous health benefits – from improving digestion and metabolism to flushing out toxins and waste. So don't forget to drink up! Your body will thank you for it.

Purge negativity from your life.

Negative thoughts and feelings can have a profound effect on our lives, leading to low self esteem, fear of failure, stress or even depression. It is essential for good mental health to rid ourselves of this negativity and take control of our own lives. Here are some simple tips to help you purge negativity from your life:

1. Recognize negative thinking patterns - Self-awareness is the first step in recognizing and managing negative thoughts. Notice how often those negative thoughts come into your mind, as well as the particular words or phrases that accompany them.

2. Challenge your assumptions - Take time to question any thought that makes you feel bad about yourself or pessimistic about a situation. Is there evidence that supports this assumption? Are there alternative interpretations?

3. Replace negative thoughts with positive ones - Make a conscious effort to replace your negative thought patterns with more optimistic and constructive ones. Instead of "I could never do that", try "I can learn how to do that".

4. Seek social support - Reaching out for help from family, friends or even professionals can be an effective way of dealing with difficult emotions. Having someone to talk to who will listen without judgement can have a calming effect and help you find better ways to deal with your negative feelings.

5. Exercise regularly - Research has shown that physical activity can boost mood and reduce stress levels, so make sure you carve out time in your day for some movement. Even just a brisk walk outdoors can make a difference.

6. Practice gratitude - Cultivating a sense of thankfulness for the people and experiences in your life can help you appreciate the good things around you, rather than dwelling on what's not going well. Consider writing down 3-5 positive thoughts each day before bed or when you wake up to start your day off with an attitude of gratitude.

These are just a few tips that can help you purge negativity from your life and take back control over how you feel. Make a conscious effort today to identify and challenge any negative thought patterns so that you can experience more happiness and contentment in your life!

Printed in the United States
by Baker & Taylor Publisher Services